UNIVERSITY COLLEGE OF
AND YORK ST. JOHN

WITHDRAWN

1 0 JUN 2022

D

105044

D0257352

1 0 JUN 2012

NOTTINGHAM

PUFFIN BOOKS

THE AIR-RAID SHELTER

Big Scott Bradley is always chasing Adam. He is a year older, a lot taller and a bully if ever there was one. When Adam discovers a real secret hiding place, where he and his sister Rachel can do whatever they like on their own, he thinks he has never been happier. The air-raid shelter makes a perfect hide-out; even if it is rather dark and spooky.

It's not long before the Bradley boys sense that something unusual is going on, but rather than spoil the secret, something else happens that unites all the children, and they forge a plan to protect their new-found den.

Jeremy Strong

The Air-Raid Shelter

Illustrated by Doffy Weir

PUFFIN BOOKS

PUFFIN BOOKS

Penguin Books Ltd, 27 Wrights Lane, London w8 5TZ (Publishing and Editorial)
and Harmondsworth, Middlesex, England (Distribution and Warehouse)
Viking Penguin Inc., 40 West 23rd Street, New York, New York 10010, USA
Penguin Books Australia Ltd, Ringwood, Victoria, Australia
Penguin Books Canada Ltd, 2801 John Street, Markham, Ontario, Canada L3R 1B4
Penguin Books (NZ) Ltd, 182–190 Wairau Road, Auckland 10, New Zealand

First published by A. & C. Black (Publishers) Ltd 1986
Published in Puffin Books 1987

Copyright © Jeremy Strong, 1986
Illustrations copyright © Doffy Weir, 1986
All rights reserved

Made and printed in Great Britain by
Richard Clay Ltd, Bungay, Suffolk
Filmset in Plantin

Except in the United States of America,
this book is sold subject to the condition
that it shall not, by way of trade or otherwise,
be lent, re-sold, hired out, or otherwise circulated
without the publisher's prior consent in any form of
binding or cover other than that in which it is
published and without a similar condition
including this condition being imposed
on the subsequent purchaser

COLLEGE OF RIPON AND
YORK ST JOHN YORK CAMPUS
CLASS NO.
SCH 823.
914
STR
JC.
2364316
DATE
15·1·92

The Discovery

Adam had never run so fast, but he knew he couldn't get away, not with Scott Bradley after him. Scott was nine – a year older and a lot taller. Adam hurtled round a corner and sped down the long alley. His feet jarred on the concrete, punching the breath from his lungs. He knew he couldn't make it to the safety of his house, not even to his garden.

A yell echoed between the corrugated iron walls of the alley, bouncing down behind Adam like some huge ball of hot air.

'You four-eyed earwig! I'll get you!'

Adam caught a glimpse of Scott closing in fast, then he was out of the alley and on to the street, his street, with his house right up at the far end, just like a mirage in a hopeless, hostile desert. He still had the high corrugated wall on one side. He put his head down and gritted his teeth against the stitch that was stabbing at his left side.

He saw a loose flap of iron sticking out from the bottom of the fence and without a second thought was on his knees, pulling and tugging at the flap, bending the metal upwards. Then he pushed himself through the hole to the other side. He fell back against the fence, panting.

'I'll get you Martello!' Scott's pounding feet went straight past. Adam closed his eyes and went on panting, gradually easing up and beginning to think straight. What was all that about? Why were the Bradley kids always on at them? He hadn't done anything so far as he could remember.

He opened his eyes. His spectacles had steamed up with the heat from running. He took them off and wiped them on his shirt-tail. Then he gazed round the scrapyard he had crawled into.

It wasn't a scrapyard at all, though that was what everybody called it. Once it had been a big garden with a posh house at the top end. The house was still there, but only because it was covered in ivy which held it all together. By rights it should have collapsed from neglect ages ago.

Most of the windows were boarded up or smashed. The front door had planks nailed across it and people had sprayed messages all over it with aerosols. The garden itself was full of junk that people had heaved over the iron fence. No wonder they called it the scrapyard. Once it had blossomed with rhododendrons and tulips, climbing roses and dahlias and lupins. Now, old washing machines rusted away on the overgrown lawns, and sad mattresses slowly faded away in the flower beds.

Adam stood up. He had never been in the scrapyard before. It was a new world. He felt as if he'd been beamed down from some orbiting space station. Now, like an alien, he stepped carefully amongst the trailing plants and derelict machinery. The house stared back with sightless windows like an old, blinded monster, sitting and waiting for the end.

He didn't like the house. There was something empty and sinister about it. As he turned away a long yell floated on the air from somewhere far beyond the garden. Scott screamed out something about murder and death and lots of other words, most of which he had probably got from the graffiti on the front of the old house. Adam grinned.

A soft silence settled once more. He could hardly hear the traffic in the street, maybe the occasional horn or siren, but that was all. He wandered towards the end of the garden where there was a large copse of trees, thick with summer leaves. There was a sort of dip in the ground amongst the trees and sticking up out of all the dead leaves was a huge square lump of concrete.

Adam scrambled down into the dip and put his hands against the flat, cold sides. He pulled himself up onto the top and walked right round it, wondering what it could possibly be. Then he jumped down and went round the outside. In one wall there was a small round hole, about the size of a drainpipe. He crouched down and peered into it and saw the sort of close-up blackness you get when you shut your eyes at night. He put his mouth to the hole and shouted.

His voice boomed about inside. Then on the next side he found a small doorway. It was open and black and there were steps leading down into the darkness. Adam braced his hands against the sides of the entrance and poked his head a little way inside. He didn't dare go in any further.

'Hallo! Hallo!'

His voice sounded big and deep, echoing around inside the little building. Adam stepped back and walked round it once more. Then he got up on the roof and sat there, with his legs dangling over the entrance. Why build something like this, half buried in the ground, with great thick walls and no windows?

Maybe it was some kind of animal cage. Maybe the people up at the big house had kept snakes down here. They could have put meat on the end of a stick and poked it through the special little hole he had found on the other wall. Or maybe they'd had a pet hippo? No – why keep a hippo in the dark? It must have been some night-time creature.

Adam drew his knees up beneath his chin and thought. The only night creatures he knew of were werewolves and vampire bats and he didn't really fancy meeting either of those.

Suddenly he leaped down from the roof and scrabbled madly halfway up the bank. Of course! It was a grave, a burial chamber . . . skeletons and coffins. . . He stared back at the black mouth of the concrete chamber, whilst slowly edging his way higher up the dip. Maybe the door had been battered down from the inside by dead bodies coming to life, trailing bandages, staring out of dark eyeless sockets.

'Aaaagh!' Adam jumped out of his skin as he backed into the branch of a tree. His heart was thumping away and he pressed the palms of his hands against his chest to stop it bursting out of his body altogether.

As he got his breath back he began to think a little more clearly. He was being stupid. If it had been a grave there would have been writing about the people inside, but he hadn't seen any.

Adam hovered on the edge of the steps and tried to pluck up enough courage to explore inside. He managed to get down the first three, but then the darkness inside was so great that fear overtook him again and he made a quick retreat into the fresh air and bright sunlight of the garden.

Even so, it was a real find, this 'thing'. It would make a superb camp – much better than the one he and his sister Rachel had built out of cardboard a couple of months ago. Overnight rain had turned that one into a spongy cardboard pudding.

He scrambled to the top of the dip and went back to the gap in the fence where he had made his escape from Scott. He knelt down and stared through. A big pair of feet clumped past. It was an odd view of things. More legs flashed. There was a pair of bright orange socks, and one had a hole in the side. A rather fat pair of lady's legs stopped right by the gap. Adam could hear the legs' owner jawing away to a friend. He grinned and picked a long grass stem.

Very carefully he poked the stem through the gap and rubbed the end on one thick ankle. A few moments later a hand reached down and scratched the place. Adam whipped the stem away and waited. Then he poked it through again and tickled the ankle. Down came the hand. *Scratch, scratch.* And again. *Tickle, tickle, scratch, scratch.*

All at once the legs clumped off, leaving Adam with a clear street. He wriggled through and ran the

rest of the way home. His mother was sitting in the front room drying her hair whilst looking at a magazine. The TV was on and Rachel was hunched on the floor as usual, staring at it. The picture was zig-zagging furiously because Mrs Martello had the hair drier on and it interfered with the signal. Rachel didn't seem to mind.

Adam went to her side and peered into her face. He half expected to see little red zig-zag lines flickering across her eyeballs, but they were just staring fixedly at the TV. He waved one hand in front of her face. Rachel made a feeble effort to brush it away. Adam flopped into a chair.

'I'll get the doctor,' he said mournfully. Mrs Martello switched off the drier.

'What?'

'I'll get the doctor.' Adam nodded towards his sister. 'She's flipped again. Her eyes have got the square look. She's been hypnotised by the adverts.'

'Shut-up,' drawled Rachel, without moving.

'See? She even talks like a zombie.'

'Leave her alone,' said Mrs Martello. 'She's not doing any harm. How come you're so dirty?'

'I got chased by Scott Bradley.'

Mrs Martello have a long sigh and slowly combed through her hair. 'Why was Scott chasing you this time?' Adam shrugged. His mother said nothing, just gave a snort and switched the drier back on.

Adam got up and tapped Rachel on the shoulder. She looked up and then followed him out of the room. 'What do you want? I was watching TV.'

'You can't watch that junk, not all higgledy-piggledy like that. Your brains will get scrambled.'

'Ha ha.' Rachel was only a fraction taller than Adam, even though she was ten and he was eight. Everybody in their family seemed short.

'Listen,' said Adam. 'Scott was chasing me and I gave him the slip. There was a hole in the fence around the scrapyard and I got through.' He went on and explained what he had found. Rachel began to show some interest.

'How big is it?' she asked.

Adam patted her cheeks softly and poked her stomach. 'If you lose a bit of weight you might get through. Are you coming to have a look?'

Rachel didn't even need to be asked. She was halfway out of the house.

'Hang on,' yelled Adam. 'We'll need a torch.'

The Secret Message

'Have we got a torch, Mum?' Adam asked.

'I thought you both had torches of your own. Grandad gave them to you last Christmas.'

Adam twiddled his fingers together. 'Mine's broken. The bulb broke.' Mrs Martello looked at Rachel, who shrugged her shoulders and made wide eyes.

'I don't suppose you know where yours is?'

'Yes I do. I do know where it is.'

Mrs Martello sighed. 'Why haven't you got it then?'

'Because it's on a bus somewhere.'

'On a bus! Then you have lost it.'

'Not really,' said Rachel. 'I know I left it on the bus but I don't know where the bus is,' she added rather lamely. Adam shuffled his feet and asked if there was another torch anywhere.

'No there isn't.'

'Well, have we got any candles then?'

Mrs Martello folded her arms and leaned back against the wall. 'What do you need candles for?' So Adam had to explain about the concrete camp in the scrapyard. She managed to find two stubby candles from the back of the kitchen cupboard. Then she gave Rachel a box of matches and a warning.

'Don't do anything stupid with them.'

'As if I would!' said Rachel.

'As if you would!' repeated Mrs Martello. 'That's why I'm warning you. It sounds as if you've found an old air-raid shelter. We used to have one in our own garden when I was a girl, but it got smashed up and taken away eventually.'

'Air-raid shelter?' said Adam. 'What was it for?'

'During the Second World War, if the enemy bombers came over, the sirens would go off and everybody had to go down into the shelters to take cover.'

Adam was impressed. 'Did you ever have to take cover?' Mrs Martello laughed and brushed her hair away from her face. 'No! I'm not that old. I wasn't born until after the war was over. But your Gran and Grandad had to use it often, especially when the doodlebugs came over.'

'Doodlebugs? What were doodlebugs?'

'A sort of flying bomb, like a rocket. Just as they got over the town they would run out of fuel and you'd suddenly hear the roar of the engines stop and everything go silent. Then you knew they were about to fall out of the sky and smash down in the city somewhere.'

'Wow!' murmured Adam. Rachel tugged at his arm.

'Come on. I want to see what this air-raid shelter looks like.'

When they got out on the street Adam looked carefully all around for any signs of the Bradley brothers, but it was all clear.

'We'll have to run,' he told Rachel, 'in case they come out.'

He made a rapid dash for the pulled-up fence and slid through. Rachel was right behind. She crawled through into the garden and stood up slowly taking everything in.

'It's fantastic. Like a secret land.'

'Where's the air-raid shelter?' demanded his sister. Adam led the way to the back of the garden and down into the dip. They stood outside the small black entrance and stared down into the inky darkness. A fusty, musty smell drifted out of the doorway.

'Ugh,' muttered Rachel. 'It smells foul.'

They hovered outside the entrance. Adam picked up a stick and threw it down the steps. It clattered against a wall, and was swallowed up whole by the dark.

'Are you going in or not?' Rachel asked. Adam gripped his candle and peered through the doorway.

'I'll go in if you do,' he said.

'I don't mind going in,' declared Rachel, trying to sound as if her heart wasn't beating three times as fast as normal. 'I don't mind going in, if you go in first.'

'No, you go first,' said Adam.

'No. Boys first.'

'Girls first. You go,' said Adam.

'But you're the youngest,' said Rachel.

'You're the eldest,' said Adam.

They stood there and looked at each other with set faces.

'The thing is,' Rachel said slowly, 'I don't mind going in, but . . . but there might be something down there.'

'Like what?'

'A skeleton or something, you know, left over from the war.'

'Don't be daft,' said Adam, but he was afraid too. He poked his head into the entrance and shouted. 'Hallo! Is anybody there?' There was a long silence. 'See? It's okay.'

'A skeleton couldn't answer,' Rachel pointed out. Adam grunted and snatched the matches from his sister. He lit his candle and started down the steps.

'Hang on,' shouted his sister. 'I'm coming. These steps are slippery.'

There weren't many steps and they soon found themselves in a small square room with a very low ceiling. It stank of damp and their feet splashed in little pools of water. The two candles threw flickering light all around, and their own shadows trembled black and twice their own size on the dripping walls.

'It's weird,' whispered Adam.

'There could be ghosts,' murmured Rachel. Adam gave a snort. 'Or crocodiles,' she added.

Adam gave a little squeal and hurriedly searched the floor.

'Do you really think so?'

'Adam, do you honestly think you'd find crocodiles in an air-raid shelter? Why do you have to think there's always something awful wherever you go? You can see there's nothing here.'

Adam felt foolish. 'Well, there are spiders down here anyway, so there. And there's writing too. Look, on the wall up there. Somebody's carved something. Maybe it's a message, a secret message from the war left by a spy. Hold your candle higher. What does it say?'

Rachel lifted her candle and craned her neck to see the sign. The writing was half blotted out by bit spots of mould. She screwed up her eyes and worked out the scrawl. '*Robert loves Eileen*. Wow! Some secret message! Those Second World War spies must have thought that was incredibly top secret.'

But Adam took no notice because he had found some more writing scratched on the wall. 'This one says *R. E. Tomlinson, 16th June, 1941*. R was for Robert, I expect,' guessed Adam. 'I wonder what happened to him. He might have been killed by one of those didgeridoos.'

Rachel groaned. 'You're an idiot,' she said. 'They weren't didgeridoos. That's a kind of musical instrument. You mean doodlebugs. You blow down didgeridoos.'

They studied the shelter more carefully. Now that they had got used to it the place seemed quite homely. It wasn't very large, but there was plenty of room to put in a table and chairs. And the beauty of it was that it was so hidden away, a real discovery. Adam smiled.

'It's really great. The Bradleys will never think of looking down here for us and we can come whenever we like.'

Rachel grinned back. She waved her candle round the room and looked about with a measuring eye. 'We ought to bring some things here and make it into a real home. We could sweep out the puddles and put a couple of old wooden boxes down to sit on.'

'We could bring the camping table down here and, hey! Dad's got some old paint in the cupboard under the stairs. We could paint the walls.'

They gazed at each other with huge, candle-lit eyes, then dashed out of the shelter and raced home to make a start on everything at once. As they reached their gate a yell came from the other side of the street.

'Four eyes!'

It was Scott Bradley, leaning out of his bedroom window and making rude signs. Adam stopped and looked back from the safety of his own garden.

'Spongebrain!' he cried, and hurried indoors before Scott could reply.

3

Adam gets the Pox

They couldn't ask their mother about the paint. She was on the telephone. Her face was red and she was standing stiff and straight, with a hand on one hip.

'Don't you talk to me like that, Beryl Bradley. Adam's only eight and your oaf is not only nine years old but he's built like Frankenstein as well . . . yes, I did say Frankenstein. He's a monster, your son, and you know it . . . How dare you? The number of times he's been caught doing . . . Huh! I don't have to listen to insults from you!'

Mrs Martello smashed the phone down and turned on Adam and Rachel. 'Don't you even speak to those Bradleys,' she warned, wagging one finger as if she wished it would wipe the Bradley family from the face of the earth.

'They're nothing but trouble. Just because they've got two video recorders they think they're Lord and Lady Wonderful. Don't you go near them.'

Adam had no intention of going anywhere near them and just nodded dumbly. He also had no intention of asking his mother about the paint, not when she was in that mood. But Rachel smiled brightly and tried to change the subject.

'Mum? You know all that spare paint under the stairs?'

Mrs Martello's eyes narrowed and she clamped both arms across her chest. 'They're a disgrace to the street, those Bradleys.'

'Mum? Can we have the paint under the stairs?'

'They've absolutely no idea how to behave, no manners and no consideration for others. They'd walk all over you if you let them . . .'

'Mum? There's some old paint tins under the stairs. Can we have them?'

'What? Yes. They ought to be taught a lesson or two . . .'

Mrs Martello stood there glaring into space. Rachel took Adam by the arm and pulled him away from the telephone. He was always amazed at the

way Rachel handled their mother. When she was in an explosive mood he would rather be a million miles away. She was likely to blow up at any moment. That's what usually happened when *he* tried to talk to her when she was in such a temper.

Somehow, with Rachel it was different. Rachel would just go on talking about what she wanted and Mum answered her as if in a trance. There really wasn't much fairness in life at all.

'Come on then. Mum said we could have the paint.'

Adam let out a long sigh and trailed after her to the cupboard. There was a pile of paint tins at the back. Rachel scrabbled about amongst the old push-chair and the ironing table and the step-ladder and everything else that had somehow got shoved into

the cupboard and forgotten.

'What shall we have? How about white walls and a red ceiling?'

'I don't think Mum will be very pleased,' murmured Adam.

'Don't be daft. She said, didn't she?'

'But she wasn't listening really,' Adam said. 'She won't be pleased when she finds out.'

Rachel had already pulled out the right tins. 'Ugh. It's filthy in there. I saw a spider as big as a football, hiding in a corner.'

That was another thing about Rachel. Spiders made Adam's knees tremble, but Rachel talked about them as if they were her best friends.

'It can't have been that big,' muttered Adam.

They managed to find a couple of old paint brushes, then they crept outside. A quick glance at the Bradley's house across the road showed that it was safe for the moment, and they walked rapidly down the street to the corrugated iron fencing. A moment later they were safely inside the scrapyard.

The silence closed around them. It really was another world, another place, almost another time, where nothing from beyond the fence could get at them. Adam drew a deep breath and grinned. He ran ahead of Rachel to the shelter. He wondered if it would still be there. Perhaps it was simply part of a dream that would burst and vanish.

But the shelter was still there, big and chunky, hard as concrete. Adam touched it with the palm of his hand. It was cold and rough and slightly damp. He patted the surface, then pulled out his candle stump and lit it, leading the way down the steps.

Rachel prised the lids off the tins. 'You do the ceiling and I'll start on the walls.'

'But you're taller than I am,' Adam moaned.

'Stand on that box then.' Rachel already had a brushful of white paint and was slopping it over one wall. Adam started on the ceiling.

It wasn't long before his arms were aching madly. The candles didn't give off much light either and it was quite a strain staring up at the dark, mouldy ceiling all the time. After a while Adam began to see spots in front of his eyes. He closed them and shook his head. When he opened them the spots were still there – great red floating spots right in front of him. There was something wrong with him. He was sickening for something.

'I can see spots in front of me,' he said anxiously.

'Just keep painting,' said Rachel. 'Stop moaning.'

Adam stood on the box, his brush dangling loosely from one hand. He looked across at his big sister. 'They're still there. They won't go away. My eyes have gone all blurry.'

Rachel sighed and put her brush down. 'Get down,' she ordered. 'Come over here.' Adam jumped off the box and Rachel turned his face towards the light of one candle. She began to giggle.

'You've got red paint all over your glasses. No wonder you can see spots.'

Adam whipped off his spectacles and the spots vanished, leaving him feeling idiotic. 'Well you'd have spots too if you'd been doing the ceiling,' he shouted.

'You've got the pox, that's what it looks like. You've got spots all over your face. You look a right mess. Come on, give me the brush. You do the walls for a bit.'

They swapped places and Adam tried to rub the paint from his glasses, but he only managed to smear it round even more. They worked on in silence. It took another half hour to complete the work, but it was worthwhile. The candle-light bounced off the newly white walls and made the place shine.

'It's fantastic,' breathed Rachel. 'It's a real home. We could get a small fire going and cook things down here. She glanced at her watch. 'It's past tea-time. We'd better get back.'

It wasn't until they got out in the open air that they both saw just how much paint they had managed to slop all over themselves. Adam had red spots across his face and hands. Rachel had the hands of a ghost.

The street was still clear as they walked home. Adam pointed out that the Bradleys were probably having tea, or watching two videos at the same time.

'I think it's stupid to have two videos,' said Rachel. 'It's just being greedy.'

They forgot about videos as soon as they got inside. When Mrs Martello saw them she went straight into her impression of an earthquake, complete with tidal waves, major flooding and hurricane force winds.

'What have you two been doing?' she screamed. 'Look at you Adam, you look as if you've just crawled out of a car crash. Where did you get the paint from? What have you been doing?'

'We did ask if . . .' began Rachel.

'Don't even speak to me,' screeched Mrs Martello. 'Get upstairs and run a bath and I don't want to see either of you until you're clean.'

Adam trailed after Rachel to the bathroom.

'I said she wouldn't be pleased,' he pointed out.

'Oh, shut-up!' Rachel snapped. 'Spotty!'

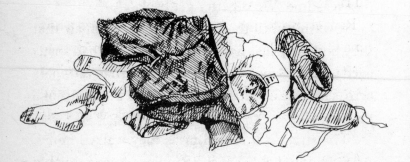

4

The Secret Weapon

The next few days saw a lot of changes in the shelter. The floor was swept out and washed down. Rachel found an old tin at home and filled it with biscuits and some apples. Mrs Martello never noticed the biscuits but she did wonder about the apples.

'That's at least five gone in one day. Who's been eating them?'

'Nobody,' said Rachel, which was quite true. They were still lying untouched in her tin.

'Not me,' Adam said.

'You'll get stomach ache.' Obviously their mother didn't believe either of them. 'And it will serve you right.'

Rachel also bought some more candles and a box of matches. Adam used some of his pocket money to buy an aerosol can of air freshener. It was called Summer Fields and had a picture of poppies and cornflowers on the label. Rachel was very rude about it.

'What did you get that for?'

'To make the shelter smell better. It smells horrible with all that old mud and damp and dust and things.'

He took the cap off and gave the shelter a good spray. Rachel clutched her throat and began to choke. She staggered up the steps and lay on the grass outside, writhing about with a cross-eyed look of extreme agony on her face. Adam came out after her, a bit red in the face and with his eyes watering.

'It's not that bad,' he insisted.

'It's awful,' Rachel moaned. 'It smells foul. Summer Fields! It smells more like Summer Dustbins if you ask me, and it stings, too.'

'I sprayed a bit too much, that's all. It's only a small room. It will have settled now.'

When they went back inside it wasn't so bad, and even Rachel admitted the air smelled a bit sweeter. They sat down by the table. Rachel opened the tin and they had some biscuits and an apple. For a while they just sat and ate and looked around at their handiwork.

'What shall we do next?' asked Adam. 'We've painted the walls and we've got chairs and a table and food.'

'What about beds,' suggested Rachel.

'Beds? Where would we get beds from?'

'There's at least one lying in the garden, and a mattress too.'

'I'm not lying on that mouldy mattress!' cried Adam. 'It's all wet and half the stuffing is falling out. It's probably full of slugs and earwigs and snails

and things. Anyway, Mum and Dad would never let us sleep down here.'

In actual fact Adam didn't at all fancy spending a night in the shelter. During the day it was all right, but the dark made a lot of difference. Rachel grinned at him.

'You're scared,' she said.

'No I'm not. What if I am anyway? I bet you'd be scared too.'

Rachel leaned forward and puffed out the candle. Suddenly it all went black and she began to laugh – a deep, hollow laugh.

'Huh huh huh huh, I'm coming to get you Adam. This is the Two Headed Bogeyman and I'm coming to get you. Slurp, slurp!'

Adam backed away towards the steps. 'Pack it in Rachel. I'm not scared,' he said, trembling from the toes upwards. 'Anyhow, I know what kills Bogeymen.'

'Huh huh huh, what's that, my little podgy supper?'

'Air freshener!'

He sprayed a short burst of Summer Fields and fled up the steps. Rachel came yelling after him and the two of them plunged out through the fence and on to the street, laughing and clutching at each other. They staggered along the street towards home and all at once came face to face with the Bradley children.

Scott stood there, his feet planted wide apart, his arms folded, a mean grin on his big round face.

Wayne was there too, looking rather worried and half hiding behind his big brother. Even four-year-old Piggy was there, standing right by Scott and copying him exactly, even though he was half the size of Adam. Darren was his proper name, but everybody called him Piggy because of his squashed up nose and the little round belly that hung over his waistband.

'Hallo, hallo, hallo!' Scott grinned. He rolled some gum round his mouth and blew a large bubble. It popped.

'Hallo hallo hallo,' echoed little Piggy.

Adam and Rachel stood quite still, a few steps from the three boys. Adam was rapidly trying to work out in his head if there was any chance of escaping back to the safety of the scrapyard. He reckoned they had a chance. He nudged Rachel with one elbow and whispered from the corner of his mouth.

'Run for the camp.'

'No,' hissed Rachel. 'If we go back there they'll follow us and find it.' She faced Scott. 'Excuse me. We'd like to get home.'

Scott burst out laughing and prodded Wayne. 'Hey! Wayne! Did you hear that? *Excuse me!* You are posh, aren't you?'

Rachel turned a deep pink colour.

'You're posh!' shouted Piggy, with great enthusiasm.

'You tell 'em, Piggy,' laughed Scott.

Adam clenched his fingers up tight into the palms of his hands. He kept telling himself that it was wrong to hit people wearing glasses, so maybe he'd be all right. On the other hand, he was sure there was going to be a fight, and he didn't know where to put his thumbs. A friend at school had told him once, but he always forgot. He couldn't remember if you stuck them down inside your fists or left them bunched up on top. He nudged Rachel again.

'Where do you put your thumbs?' he whispered hoarsely.

'What?'

'Where do you put your thumbs?'

Rachel glanced at her brother, certain that he

must have gone mad, but what he said made her look down at her own thumbs and she saw the tin of air freshener. She had to bite her lip to stop a cunning smile spreading from ear to ear. 'Get ready to run back to the camp,' she muttered.

'What are you going to do?'

'You'll see.' She went on in a loud voice. 'Are you going to let us pass or not? Because if you don't then we'll just have to use our secret weapon, so you'd better watch out.'

Scott stopped grinning and Wayne slithered even further behind his big brother. 'What secret weapon?' Scott demanded. Rachel held up the tin, keeping the label covered over. 'What's that?' asked Scott.

'Airoooooff!' Adam gasped as Rachel jammed an elbow into his ribs.

'Paint,' she said quickly. 'Bright green paint, and if you come a step nearer I shall spray your legs bright green.'

'Oh yea?' said Scott. He blew another bubble, rather unsure what to do about this new threat.

'Oh yeah!' shouted Piggy.

'Oh yeah,' said Rachel quietly, and she and Scott stared at each other long and hard, like two gun-slingers out on a dusty road.

Poor Scott was in a fix. If he'd been by himself he would have backed down and gone home. He was scared of the tin. But he had Wayne and Piggy with him and he didn't want to be made to look like a cowardly fool in front of his younger brothers. He hunched up his shoulders.

'We're not moving,' he said.

'We're not moving!' yelled little Piggy, trying to hunch his shoulders too.

Rachel's heart was racing into fifth gear. 'Get ready,' she hissed at Adam. 'Okay then, you asked for it!' She darted forward, spraying down at Scott's legs and yelling to Adam at the same time. He turned and fled back to the scrapyard. He didn't see Scott lean backwards with a startled yelp, expecting bright green paint to fly all over his legs. Scott crashed into Wayne and they both fell in an angry confusion of legs and arms.

Rachel raced away after Adam and slipped through the hole in the fence. Scott was on his feet in an instant, realising he'd been tricked. His anger made him run twice as fast after Rachel and he was just in time to see her disappear through the hole. He dashed after her.

Adam was already safely in the air-raid shelter when Rachel came crashing down the steps, panting madly. 'He saw me,' she gasped. 'He's after us!'

They heard a yell outside the shelter. 'I know you're in there. You can't get away now. I've got you trapped!'

There was a short silence, during which Adam and Rachel could only hear some mutterings as Scott gave instructions to Wayne and Piggy. Then Scott started up again.

'I'm going to sit down right here by the door,' he said. 'You'll have to come out sooner or later.'

5

Peace Breaks Out

Rachel lit a candle. The little flame flickered, spluttered and rose straight. It cast strange shadows across Adam's face, making his worried frown look like the snarl of an evil magician.

'Now what do we do?' he asked. Rachel fiddled with the box of matches. She didn't answer immediately, but eventually said, 'They won't dare come down here.'

'Why not?'

'Because the steps are too dark and narrow. We can get them easily.'

'What with?' asked Adam, wondering how they could attack the Bradley brothers with only two wooden boxes and a table.

'It doesn't matter,' Rachel snapped. 'They don't know that we haven't got any weapons.'

Adam could see the sense in that. In fact it sounded pretty good. He went to the bottom of the steps and yelled up them. 'You'd better not come down here! We've got a rifle and a knife and a World War Two hand grenade and some dynamite . . . ow!' Rachel pulled him away from the steps.

'Shut-up you idiot. What on earth are you going on about?'

'I was making them think we're well armed.'

Before Rachel could say any more, Scott shouted down the steps.

'We don't care if you've got a bomb down there – doesn't make any difference to us. We'll stay here until you have to come out. Anyway, what was that stuff you sprayed on my legs? It pongs like mad.'

Adam was about to tell Scott it was only air freshener, but Rachel stopped him. 'It's pesticide,' she lied. I expect your legs will fall off soon.'

'Oh yeah?' said Scott, disbelievingly.

'Yeah,' parroted Rachel.

There was a pause and then they heard Scott ask his middle brother if there was anything he could wipe his leg with. Wayne went off to look. Scott stayed right outside the entrance, casually swishing a big stick in the air and knocking the heads off nearby dandelions and nettles.

Rachel sat down and looked across at her brother. She felt cold and there was a tight sensation inside her stomach that made her feel uncomfortable all over. 'What shall we do?' she whispered.

Adam leaned forward. 'Listen. I've been thinking. Suppose we let Scott and his brothers come down here?'

'Great. Then what do we do? Crawl out through that bit of drain pipe?'

'You don't understand,' said Adam. 'Let them

come down here and we'll share the camp with them.'

'You're a loony. Why should we do that?'

'Because I don't want to starve to death and because I don't want to get beaten up and because the shelter is big enough for all of us and we could have better games and everything if there were more of us . . .'

Rachel sat and thought. Adam started to say more but she told him to be quiet because she was thinking. At last she stood up.

'Okay. We'll try it.' She went up the first few steps. 'Listen, are you still out there?'

'Yeah,' drawled Scott.

'Do you know what this place is?'

'Yeah,' said Scott again. 'It's your coffin.' Wayne giggled.

'It's an air-raid shelter from the Second World War,' Rachel went on. 'It's our secret camp. We've painted the walls.'

'And the ceiling,' added Adam, remembering the red spots.

'We've got candles and food,' said Rachel. 'You can share it with us if you want to come down.'

There was a very long wait as Scott slowly stood up and came to the top of the stairs. He stared down into the gloom. Wayne hovered behind him and Piggy stuck his head between Scott's legs and tried to see down the stairwell. Scott was uncertain.

'It's a trick,' he muttered to the others.

Rachel shouted up the stairs again. 'Are you still there? You can come down if you want to, but only if you come in peace.' Scott shifted his feet and almost squeezed Piggy's head flat, forgetting where it was. Piggy began to cry quietly.

'What about the rifle and the hand grenade and everything?' Scott asked. 'Throw them out first, then we'll come down.'

Rachel sighed. 'We haven't got any.'

'Oh yeah? You said you had. And what about the pesticide?'

'That was only air freshener. That's why it smells. Do you really think I would have sprayed real pesticide on your leg?'

Scott didn't want to answer that. It made him uncomfortable to discover that there were people who wouldn't even do that when they were in a spot of bother.

'Promise it's not a trick,' he demanded.

'Promise you won't start a fight,' demanded Rachel.

'Fight fight!' yelled Little Piggy, thinking the Bradleys were going into battle yet again. Scott turned on him. 'Oh shut-up, you bonehead!' Piggy was so astonished that for a second all he could do was stare up at his hero-brother with pure shock. His mouth slowly turned down at the corners and he was on the point of bursting into a flood of tears when Scott grabbed his hand.

'Come on. We're going down into their camp. Hold tight, Piggy. We're coming down!' he shouted.

They stood inside the shelter and gazed around. Adam and Rachel pressed rather fearfully back against the walls, wondering what they had let themselves in for. The little candle fluttered madly in the draft. What seemed like hours passed as they just stood there, slowly looking round at everything – the painted walls, the ceiling, the candle, chairs and table.

Scott's mouth hung open. He forgot to chew his gum. Wayne slowly turned his head this way and that. Piggy gazed with shining eyes at the glittering candle. He held out a hand to it. 'It's a star,' he cried.

Rachel laughed. 'It's a candle,' she said and she bent down and puffed it out, to prove it. Instantly there was chaos.

'Star's gone! It's dark! I want to go home!' squealed Piggy.

'Light the candle,' Scott snapped. 'Shut-up Piggy!'

There was a scuffle as both Adam and Rachel searched for the matches in the dark, then the candle was alight once more.

'It's pretty,' Piggy sighed, the smile returning to his face.

'It's a real camp,' breathed Scott. 'It's fantastic. A real World War Three camp.'

'World War Two,' corrected Adam. 'We haven't had World War Three yet.'

'Hey, Wayne, isn't it great?' Wayne nodded. He always agreed with anything his big brother said. He'd learnt that it was best to keep quiet.

Rachel got out the biscuit tin and passed it round. Adam began to feel a bit safer seeing how pleased Scott was with the shelter.

'During the war people had to come down here and take cover in case any lunarslugs landed on them,' he said.

'What were lunarslugs,' asked Scott.

'They were incredible things with rockets and when they ran out of fuel they came down and blew everything up.'

'Don't be stupid,' grunted Scott. 'Slugs don't have rockets.'

Rachel was almost choking on her biscuit, trying to butt in. 'He means doodlebugs. Honestly! Last time it was didgeridoos and now it's lunarslugs. Doodlebugs, Adam, doodlebugs.'

Adam shrugged. 'I think lunarslugs sounds much more interesting,' he said quietly. The biscuit tin was passed round again and they sat down at the table and ate. Scott said that he had tried to make a camp once in their back garden, but when the dust-men came they took it away thinking it was rubbish.

'Anyway,' he added, 'my mum likes to keep the garden neat. We're not allowed to play in it much.'

'Our mum doesn't like yours,' said Rachel. 'She thinks your mum is stuck up.'

Scott looked across at Rachel with his steel-blue eyes glittering in the candlelight. He started laughing. 'My mum doesn't call yours Mrs Martello. She calls her Mrs Marshmallow.'

Adam and Rachel glanced at each other as Scott went on. 'I don't know why your mum doesn't like ours.'

'It's because you've got two video recorders,' said Adam.

'No we haven't,' said Wayne. Everybody looked at him in surprise. It was the first thing he'd said for half an hour. 'Well we haven't,' said Wayne, going bright red.

'We've only got one,' said Scott. 'The other one we were looking after for my uncle. He's got it back now.'

'We haven't got one at all,' Rachel murmured. Scott shrugged his shoulders.

'It's nothing special, only more TV, that's all. This camp is better. It's real. Can I have another biscuit?'

Adam passed the tin round and it came back empty.

'Don't worry,' said Scott. 'We'll bring food tomorrow, won't we Piggy? It's a fantastic camp. Come on you two, we'd better get back.' The Bradley brothers tramped up the steps.

'Meet you here tomorrow,' Scott shouted down the stairs. Then they were gone.

After a while Rachel blew out the candle and she and Adam went slowly up the steps and out into the evening air.

'Scott's okay,' said Rachel thoughtfully. 'He's okay really.'

'Oh yeah, Scott's okay. It's Piggy who's the worst,' Adam replied.

'Piggy? What's wrong with Piggy?'

'He ate all the biscuits. Every single one,' said Adam.

6

Fire! Fire!

'Hey, look at that,' said Adam. They were walking down to the hole in the fence. He stopped and stared up at something stuck on the fence. Rachel came back to read it.

'*Private. Keep out.* I don't remember that, do you? Anyway, we're not doing any harm,' she said.

'There's another one here,' Adam pointed out. 'It says *Trespassers will be prosecuted.* What does that mean?'

Rachel walked on. 'Hurry up, or Scott will get there before us.'

'But what does *Trespassers will be prosecuted* mean? Are we trespassers?' Adam scrambled through the hole behind his big sister.

'Of course not. We're too small.' Adam screwed up his face. That didn't make sense. He knew he was small for his age but he didn't think it made that sort of difference.

'What about prosecuted? What about that?'

'It means that if you are a trespasser and the police catch you then they fine you.'

Adam swallowed hard. That didn't sound very nice. He didn't even have any pocket money saved up. He'd spent it all on air freshener. Did that mean he'd have to go to prison?

'Look,' said Rachel, seeing Adam's worried face. 'It doesn't matter. We're too young to be trespassers and anyway, we can always pretend we can't read.'

They went down into the shelter and lit the candle. Rachel gave a little shiver. The trouble with being so deep underground was that it was always dark and cold. She was glad she wasn't a mole. Rachel thought moles must be very cold animals, digging away underground all the time. It was probably frustrating too. Suppose you were tunnelling along at great speed and then ran slap bang into the foundations of a wall or a tree? It was little wonder moles had flat noses.

There were voices outside and the Bradley brothers came charging down the steps. 'Elephants,' murmured Rachel privately.

'Hey, have you seen those notices!' shouted Scott. 'Who put them up? Bloomin' cheek! I tried to tear them down but the paste is too strong. Who says we've got to keep out? And who's Bill Posting?'

Rachel shook her head. 'Don't know. Why?'

''Cos there are all these notices saying *No bill posting*. I wondered who he was. Doesn't make sense, does it?'

'Bill Posting!' cried Little Piggy. 'I want a biscuit.'

'Maybe there's going to be an election,' suggested Rachel, 'and the other side is saying don't vote for Bill Posting.'

Adam shook his head firmly. 'It's not that. It just means you mustn't stick notices on the fence. You know, don't post bills. They're called bill posters, aren't they?'

Scott and Rachel stared at Adam, while Piggy hunted for the biscuit tin. 'Don't be daft,' said Scott at length. 'That's just what they've gone and done, isn't it? Stuck posters up all over the place. How can you say don't do it and then do it?'

'Well, that's what it means,' said Adam staunchly.

'Biscuits!' screeched Piggy.

'All right, all right, keep your hair on Piggy,' said Scott, and he pulled half a packet of biscuits from his trouser pocket. 'Not that he's got much left,' laughed Scott, rubbing his hand over Piggy's stubbly head. 'Dad gave him a crew cut last night.'

'I thought he'd fallen under a lawn mower,' said Adam.

'Yeah, well, you've got funny eyes, haven't you Goggles?' Evidently it was all right for Scott to make jokes about his brothers but not for Adam or Rachel to do the same. Adam decided that Scott was in a bad mood. He picked up a bit of twig and idly waved it through the candle flame. It went black where the soot stuck on the wood. Adam looked at the twig and smudged the dark soot with his finger.

He stood up and held the candle flame to the ceiling. A patch of black soot quickly appeared and Adam began to write with the candle. First of all he wrote *Private. Keep out.* The others watched him in thoughtful silence. Then Adam put *Trespassers will be* . . . and he couldn't think what to do with trespassers.

Scott jumped up. 'Here, give it to me.' He snatched the candle from Adam. 'Ow, do you have to spill hot wax all over me?'

Adam grunted. He wasn't going to argue with Scott – he was far too big. Scott wrote quickly. *Trespassers will be hung, drawn and quartered.* 'That's what they used to do in the old days,' he explained. 'They'd hang them up and then they took them down and pulled out their insides and cut them into four bits.'

'Ugh,' said Rachel, and they all became silent as they considered the awful results of trespassing in their shelter. Rachel at last took the candle and wrote *Home sweet home* in careful lettering.

'Can I have a go?' asked Wayne. Scott jumped back in surprise.

'Wow! He spoke. He does that sometimes, you know.'

Wayne began to go red and Adam felt sorry for him, always being made fun of by his elder brother. 'Let Wayne have a go,' he said. So Wayne took the candle with a big grin and held it to the ceiling. Bit by bit, wobbly black writing appeared. *Rachel . . .*

'Go on, go on!' shouted Scott, jumping up and down. Wayne went on.

Rachel loves . . .

'Stop him,' said Rachel. 'That's not fair. Wayne . . .' She tried to get to him but Scott held her back.

'Go on!' yelled Scott. 'That's great! Go on!'

Rachel loves Scott.

There was a bellow of rage and the shelter was plunged into darkness as Wayne dropped the candle and fled giggling up the stairs.

'Wayne! You come back here!' Scott tore after his brother.

Adam sat in the corner laughing until tears rolled down his cheeks. Rachel picked up the candle, lit it and carefully blotted out the last message. 'I don't think it's funny,' she said.

'It is, it is,' spluttered Adam.

'You should have seen Scott's face. He thought it was going to be somebody else and it would be really funny and he held you back and everything and then, and then, then it was him! You should have seen his face.'

'You're pathetic,' said Rachel icily, and she went outside.

Wayne was nowhere to be seen. Scott came back, his face red and angry. He glared at Rachel as if she were to blame for everything, and clambered on top of the shelter. 'When I catch him I'll kill him.'

Rachel glanced up coldly. 'I don't suppose it will be the first time.'

'Clever, clever,' hissed Scott.

There was a long silence until Adam and Piggy came out from the shelter. They were quite cheerful because they'd finished off the biscuits together and Adam had helped Piggy write *Piggy loves biscuits* on the ceiling. Now they wanted to know what everybody was going to do. Rachel rubbed her arms and said they ought to do something about the cold.

Scott suggested making a fire down there, so they hunted around for dry twigs. Piggy even found some good thick branches, but Adam thought the fire might get too big if they burned those. They took their bundles down into the shelter and moved the table and boxes away from the wall where they wanted to put the fire.

Scott took over. He carefully set up the twigs, laying them this way and that. He crumpled up a little newspaper and stuffed it underneath. Then he pressed the twigs down with his hands so that they lay close together. The others watched, fascinated.

He put a match to the paper and it quickly took flame. The twigs began to crackle and little flames sparkled. They all crowded round, holding out their hands to the little fire as it struggled to get going. One or two bigger flames made their shadows dance on the white walls.

'It's lovely,' murmured Rachel.

The twigs snapped and spat as the flames grew stronger. Wisps of smoke spiralled into the air. Piggy wiped his eyes, which seemed to be watering a little. The flames leaped up and everybody took a step back. More smoke pushed into the air and hunted round for somewhere to go, but there was nowhere to go. There was no chimney and the room was small and low-ceilinged.

Piggy and Rachel began to cough. Eyes were beginning to stream.

'It's a good fire,' choked Scott, taking deep gulps of air and blinking at a hundred miles an hour.

The fire was going really well and the room was as warm as toast. It was also bursting with thick grey smoke. Rachel staggered to the steps and groped her way up into the open air. A moment later Piggy, Adam and Scott plunged out through the doorway and threw themselves on the grass, where they lay coughing and choking and laughing.

'Lovely fire!' Scott snorted. 'Look at it!' Huge clouds billowed out from the entrance and a thick twist of smoke plumed from the little round air hole.

'I think we might need a chimney,' murmured Adam. 'Anyway, we shall have to wait until after lunch for the smoke to clear.'

'It looks like a doodlebug has just scored a direct hit on us,' said Rachel. She began to scramble up the dip. 'Come on, we ought to get home.'

'You coming back this afternoon?' Scott asked.

'We've got to go out. We'll be back tomorrow.'

'I'll come this afternoon,' said Scott. 'I'll keep guard in case any trespassers come.'

Rachel looked back from the top of the dip. 'Have you got enough rope?' she called. 'And plenty of knives and things?'

'What's that for?'

'Hanging and drawing and quartering trespassers, of course.'

But Scott wasn't so easily made fun of. He frowned darkly. 'I'm saving that for Wayne,' he muttered.

7

Ghosts

That same evening there was panic in the Martello house. Adam and Rachel had discovered what the notices on the iron fence were all about. Mr Martello had read about it in the local newspaper at tea-time.

'About time too,' he said. 'They're going to develop the scrapyard and build a supermarket there.'

Adam and Rachel froze with food halfway to their mouths. Rachel slowly lowered her jam sandwich. 'What are they going to do, Dad?'

'Knock down the big house and build a supermarket.'

'That will be nice,' said Mrs Martello. 'We could do with a supermarket. I don't like that shop on the corner. Their cheese is always stale.' Adam was leaning forward, hands gripping the table edge.

'Knock down the house? They're going to build a supermarket where the house is?'

'Yes.'

'But the garden will be all right, won't it?'

Mr Martello laughed. 'Don't be daft. They'll flatten that for a car park.' He pushed the paper to one side and went to watch TV. Adam and Rachel went upstairs like the walking dead. They couldn't believe it. Just when they thought they had found the best place ever it was all going to be destroyed for the sake of some useless supermarket.

Rachel sat on her bed with her chin cupped in her hands. She gloomily pushed one slipper around with her foot. 'What are we going to do?' she asked.

There was a very long silence as Adam racked his brains for ideas. Suppose they got up a petition? No – everybody would prefer the supermarket. What about going on hunger strike? No – their mother always said they ate too much anyway.

'I don't know,' Adam at last admitted, and he went off to his own bedroom, very depressed.

That night there was a thunder storm. It wasn't very big, but the rumbles and grumbles woke Adam up and then of course he couldn't get back to sleep. He lay in bed, staring into the darkness and wondering what the shelter would be like at night. It would be nice to feel brave enough to stay down there one night, but he didn't like that kind of strange darkness – being alone in a place he didn't know well. He'd be afraid of night-monsters: witches and ghoulies and ghosts. The old house in the garden looked very much as if it might be haunted.

Adam sat up. A haunted house! But of course! Suppose they haunted the big house? Nobody would go near it then. The builders would run away and the supermarket would never get built. Adam smiled and lay back on his pillow. It couldn't fail.

In the morning he went straight to Rachel's room and had unfolded his plan before she'd even opened her eyes. So he had to go through it all over again, Rachel's face lit up.

'That's brilliant, Adam! We can get some old sheets . . .'

'And some old chain from the scrapyard,' said Adam.

'What's that for?'

'To rattle. Ghosts always rattle chains.'

'I wouldn't know. I've never been one,' answered Rachel, and she started to laugh.

After breakfast they went down to the scrapyard. By the time Scott and his brothers showed up Adam and Rachel had found several bits of chain. They had brought a couple of sheets from home. When she saw the Bradleys coming Rachel zipped down into the shelter. Adam waited at the entrance.

'I wouldn't go down there,' he warned Scott and the others.

'Why? What's wrong?'

'There's a ghost down there.'

A spine-chilling howl drifted up the stairwell. Piggy ran screaming back to the hole in the fence. Scott turned white.

'Don't be stupid,' he said. 'There can't be. It's never been there before.'

'It must have just moved in.'

Down in the dark shelter Rachel picked up a chain and rattled it furiously. She let out another long wail. Scott was uncertain.

'Where's Rachel?' he asked suspiciously. Adam shook his head.

'She went down there,' he said. 'She hasn't come out. I think . . . I think it must have got her.'

'Whoooooooooooo!' screeched Rachel, coming up the steps and dragging the chain behind her. She stepped out of the doorway. Scott turned to flee, a look of horror on his face, and at that moment Rachel dropped the sheet, and stood there grinning at him.

He was furious at being tricked, but when Adam explained what was going to happen to their shelter if they didn't do something about it, he calmed down. Rachel said that they were all going to be ghosts, so Scott went and fetched Piggy from the street, where he was standing and howling miserably to himself.

It didn't take long to cut up the sheets into ghost outfits, but it was a much harder job to break into the old house. The doors and windows were all boarded up and it took over an hour to loosen some planks so that they could crawl through.

They found themselves in a huge room with a high ceiling. It was quite bare except for dust and bits of old newspaper and other rubbish lying on the wooden floor. It was creepy. They spoke to each other in whispers and opened doors very slowly, as if there might be a skeleton or even worse behind each one.

Out in the hall they found the staircase and Scott led the way up. The bedrooms were almost as big as the downstairs rooms, and they smelt dusty and old and tired. And still there was something weird and unsettling about the house. At last Rachel put her finger on it.

'It's all dark, because the windows are boarded up. We know it should be daytime, but it seems like night. Everything is back to front.'

Adam found a big wardrobe and he gingerly opened the door. Inside were two lonely coat-hangers. Rachel rattled them with her fingers. 'It's like a house but not like a house,' she murmured.

Adam crossed over to one of the boarded up windows. 'We can see the builders from here. One of the planks is missing. We should make this our headquarters.'

'Yeah. Then we can drift across the landing, rattling our chains and howling.' Scott grinned. 'We'd better get our gear up here. Hey, Wayne, go down and get our stuff, okay?'

Wayne went to the top of the stairs and gazed down at the silent emptiness below, with all the open doors around the hall leading to – leading to what? Wayne shook his head.

'Go on,' said Scott. 'You're not scared are you?'

'I want to be a ghost!' Piggy cried, and he went tottering down the stairs, so they all followed behind him.

They put their ghostly outfits on and walked slowly backwards and forwards at the top of the stairs. For a while, the house echoed with strange yells and the rattling of five chains as they trooped about trying to scare each other to death. Eventually Scott and Wayne ended up laughing themselves silly.

'What's so funny?' asked Adam, yanking his sheet straight.

'You lot!' cried Scott. 'You all look so stupid!'

Rachel sniffed. 'I don't see anything funny about ghosts. We're supposed to be scary, not funny.'

Scott and Wayne gradually pulled themselves together and they went back to headquarters and took the robes off. Adam wiped the sweat from his face and glanced out through the window.

'There's a man outside,' he said nervously.

'What do you mean?' asked Scott.

'I mean there's a man outside. What's the matter? Do you think I need glasses?'

'You've got glasses, stupid.'

'I know!' snapped Adam. 'And I'm telling you there's a man outside on the path. Now there's another one. They're carrying hammers and things.'

There was a rush to the window and five pairs of eyes stared out. 'It's them,' said Rachel. 'They've come already. They're going to knock the place down.'

'I want to go home,' started Piggy, his lower lip slowly getting more and more droopy at the corners.

The others began to pull their sheets back on frantically.

'Come on, come on, this is it,' said Scott, pulling the two little eye holes round so that he could see. 'Is everybody ready?' He bent down and somehow managed to shove Piggy into his sheet, although it looked as though the wailing Piggy didn't want to be ghostly at all.

'Ready,' said Rachel, whose heart was thumping like a drum.

'Ready,' said Adam, gripping his bit of chain until his knuckles went white.

'Ready,' croaked Wayne, who suddenly didn't think there was anything funny about being a ghost at all.

Down below they heard the men hammering the boards off the front door so that they could get in. The five spooks crept out onto the landing.

8

Trouble

The three men who had come to the house were nothing to do with builders, and they certainly hadn't come to knock the place down. They only had two hammers, a chisel and a big screwdriver between them. One was from the council, one was a surveyor who needed to see what work needed doing, and the third man was from the supermarket company.

It didn't take long to remove the boards from the front door. The council man, Mr MacSween, unlocked the huge padlock and they went through into the hall.

'It's been empty for years,' explained Mr MacSween, and his voice echoed round the house.

'Yes. I can see that.' The surveyor picked at a piece of loose wallpaper and pulled. A huge sheet of mouldy pattern came away from the wall. The supermarket representative, Mr Davidson, coughed and said politely that old houses were full of damp. 'It doesn't matter,' he went on. 'Since the whole place will be pulled down.'

They walked through into the front room. Mr MacSween pushed some empty tins with one foot.

'Tramps. They get in everywhere. They used to sleep here until we had the windows boarded up.'

The surveyor was in one corner, peering down at the floor. 'The boards are rotting,' he said and stamped down hard so that the floor splintered into dusty fragments.

'It really doesn't matter,' explained Mr Davidson impatiently. 'We're not buying the house to live in. We're going to knock it down. I don't care how much damp and . . .'

'Ooooooooooooooooo . . .!'

Mr Davidson wheeled round and stared at the empty doorway.

'What was that?'

'An owl,' answered the surveyor coolly.

'Oh I don't think so. It didn't sound like an owl to me,' said Mr MacSween. The surveyor gave a little smile and he regarded the other two men with half closed eyes.

'It must have been a ghost then.'

'Whoooooooooooo . . . !' A chain rattled loudly. Mr Davidson swallowed and fiddled nervously with his fingers.

'Sounds like a lavatory chain to me,' the surveyor said.

'Anyhow, it's coming from upstairs . . . So if you gentlemen are brave enough, we may as well go and see.'

He led the way out into the hall. There was some quick rushing about at the top of the stairs and a white shape drifted slowly across the landing.

'Oooooeeeeoooooeeeeoooo . . . !'

Mr Davidson went tearing back to the front door. 'I saw it, I saw it!' he squeaked. 'It was a ghost, up there on the landing.' He pointed a trembling finger at the top of the stairs. Right on cue another ghost went sailing past, arms outstretched and a chain clanking away behind.

By this time Mr Davidson was right out in the garden, leaning by a tree and panting. 'I thought it felt strangely cold when we first went inside. They say that where there are ghosts the temperature always drops.' He shuddered. 'It was vile.'

The surveyor watched him with a little smile. 'Houses like this are always cold, especially when they've been boarded up. Keeps the sun out.'

Mr Davidson shook his head. 'No. It was too cold for that.'

'It was the damp then. Come on, you don't believe in ghosts do you?'

'I didn't,' trembled Mr Davidson. 'If you'd asked me yesterday I would have said there were no such things, but I've just seen one, haven't I?'

The surveyor looked across at Mr MacSween and winked. 'No, you haven't seen one. You saw two.'

'Two!'

'Yes. There were two of them, didn't you notice? One was much shorter than the other. I thought perhaps they were brothers.'

Mr Davidson looked up. 'You mean . . . You think . . .' He stopped and stared back at the house. Colour began to flood back into his face. He went from white to pink to red. 'Do you mean we've been tricked?' He turned on his heel and strode back towards the house, with Mr MacSween and the surveyor close behind.

Just as they thought they had succeeded panic set in upstairs.

'They're coming back!' shouted Rachel. 'They're coming back!'

Adam hitched up his sheet. 'Don't worry. I'll get rid of them for good.' He swept out onto the landing.

'Oooooooo . . . !' he cried and stood at the top of the stairs with his arms stretched out menacingly.

'What on earth is that meant to be?' asked Mr MacSween loudly.

'I think it's meant to be a rather short ghost,' replied the surveyor, and he started up the stairs.

'Oooo . . . !' screeched Adam. 'Beware the ghost of ghoulie land. Beware the curse of, the curse, the curse of The Hand Of Death!' he cried desperately, waggling his fingers at the men coming up the stairs.

The surveyor started taking two steps at a time and he roared out, 'Beware the Mighty Surveyor who's coming to get you little pests!'

Chaos followed. Adam fled back to headquarters and slammed the door. The others crowded round with their backs against it as the three men came up on the other side.

'I don't think it's going to work,' panted Adam.

'They don't seem to be scared of ghosts at all.'

'It was a stupid idea,' grumbled Scott. 'Look what a mess you've got us into.'

'Huh! You didn't exactly come up with any bright ideas, did you?' shouted Rachel.

'Didn't give me chance, did you? You two think you're so bloomin' clever all the time, nobody else gets a chance, do they? You and your four-eyed twit of a brother.'

'Four-eyed twit!' yelled Piggy from under his sheet.

'Oi! Who are you calling a four-eyed twit?' demanded Mr Davidson from the other side of the door, fixing his spectacles straight. He turned to the surveyor. 'How do they know I wear glasses?'

And he threw himself at the door once more.

On the other side the five children hunched their shoulders and pushed back. 'Stupid marshmallows,' hissed Scott.

'That's it,' cried Rachel, and she walked away from the door and stood with her hands on her hips. 'Come on Adam. Leave the door alone. Scott thinks he's so strong and clever, he can keep it shut himself.' She pulled Adam away from the door.

'Cowards!' yelled Scott and he launched himself at the pair of them. They collapsed in a kicking, screaming heap on the ground.

'Fight! Fight!' squealed Piggy and he clambered on top of the heaving mound of bodies.

With only Wayne holding the door, it slid open. Strong hands picked the children from the floor and separated them. Adam and Rachel and Scott stood there panting like caged animals, their ghostly costumes torn and tattered and smudged with dirt.

'What have we here?' said the surveyor with his little smile. 'A family of ghosts, and not a very happy one at that.'

'What do you think you're playing at?' Mr Davidson snapped. 'How did you get in? Can't you read the notices? Do you know what happens to people who break into houses?'

Piggy snivelled. Wayne cringed. The three older children were silent. Mr Davidson managed to sound just like Mrs Martello and Mrs Bradley put together. Mr MacSween broke in.

'How about you?' he said to Rachel. 'What's your story? Let's have the real one.' He smiled at her gently. Rachel slowly pulled off her sheet. Then she told the story, nice and slowly, but leaving out the bit about the air-raid shelter. A sixth sense told her that there was no point in giving *everything* away.

'It's quite ridiculous,' said Mr Davidson severely. 'Everybody wants a supermarket. You'll be able to buy toys there.'

Rachel sighed and glanced at Adam, wondering how long it would be before Mr Davidson understood that the scrapyard was worth a million supermarkets full of toys.

'We shall have to take them to their parents,' Mr Davidson went on. 'We can't let them get away with it.'

As they marched off down the street Scott kept calling Rachel a rat for giving the game away. Before she could tell him to shut up, the surveyor did it for her.

'I'd be quiet if I were you. If this girl hadn't told us the truth you'd be in far deeper trouble now, believe me.' Scott scowled and went silent.

Mrs Martello was furious when she heard the tale and it was made worse by the Bradley children being involved. She went red at having to listen on her own doorstep to what her own children had been up to. It was the same over the road at the Bradley's.

As soon as the men had gone, the children were bawled out.

Adam and Rachel got sent upstairs to their rooms without any tea. Scott got a clip round the ear and then Beryl Bradley was straight on the 'phone to Mrs Martello. Adam and Rachel could hear their mother yelling back down the line even from their bedrooms with the doors shut.

When their father came home, Mrs Martello told him the whole story. She was still fuming but Mr Martello just burst out laughing. He thought the whole thing was hilarious and he made Adam and Rachel come downstairs and tell him their version of events. When he heard how scared Mr Davidson had been Mr Martello laughed so much he slithered right down his chair on to the floor.

Mrs Martello couldn't keep a straight face watching her husband in that state. She began giggling too and then all four of them rolled about. So Adam and Rachel were allowed to have tea after all.

The next day they went back to the air-raid shelter. There was no sign of the men anywhere, but Scott, Piggy and Wayne were already there.

'I've got some more biscuits,' said Scott. 'And guess what? I heard one of those men talking to my Dad last night. He said it will be at least a year before they start work here.'

'A year?' echoed Adam.

'That's what he said. Have a biscuit.'

'Me first!' shouted Piggy.

'You first,' said Scott cheerfully. 'Little Piggy first.'

'But that's great,' Adam said. 'That gives us a whole year to think up a way of stopping them and saving the air-raid shelter.'

Scott took another biscuit and waved it over the candle to see if it would toast well. 'I'll think of something next time,' he said.

'It had better be good.'

'Oh it will be. It will be really, really good. And then we'll have this place all to ourselves.' Scott toasted the other side. 'It will be a really ace idea.' He nodded emphatically and crunched on his biscuit. 'Ugh!' Scott leaned over the table and spat out the crumbs. 'That was foul! Ugh!'

'Let's hope it's a better idea than making toast over candles,' said Adam pointedly, sticking his own piece down the back of the box he was sitting on. Wayne started to giggle, despite Scott's ugly frown. Soon the whole shelter was ringing with their hysterical laughter.

Some other Young Puffins

RADIO ALERT
RADIO DETECTIVE

John Escott

Two exciting stories centred on a local radio station, Roundbay Radio. In each book there's a mystery that the children involved help to solve brilliantly.

FIONA FINDS HER TONGUE

Diana Hendry

At home Fiona is a chatterbox, but whenever she goes out she just won't say a word. How she overcomes her shyness and 'finds her tongue' is told in this charming story.

THE THREE AND MANY WISHES
OF JASON REID

Hazel Hutchins

Jason is eleven and a very good thinker, so when he is granted three wishes, he is very wary indeed. After all, he knows the tangles that happen in fairy stories!

DINNER LADIES DON'T COUNT

Bernard Ashley

Two stories set in a school. Jason has to prove he didn't take Donna's birthday cards, and Linda tells a lie about why she can't go on the school outing.

MOULDY'S ORPHAN

Gillian Avery

Mouldy dreams of adopting an orphan, but when she brings one home to her crowded cottage, Mum and Dad aren't pleased at all.

THE TALE OF GREYFRIARS BOBBY

Virginia Derwent

A specially retold version for younger readers, of the true story of a scruffy Skye terrier who was faithful to his master even in death.

ONE NIL

Tony Bradman

Dave Brown is mad about football, and when he learns that the England squad are to train at the local City ground, he thinks up a brilliant plan to overcome his parents' objections and gets to the ground to see them. A very amusing story.

THE GHOST AT No. 13

Gyles Brandreth

Hamlet Brown's sister, Susan, is just too perfect. Everything she does is praised, and Hamlet is in despair – until the ghost comes to stay for a holiday and helps him find an exciting idea for his school project.

ZOZU THE ROBOT

Diana Carter

Rufus and Sarah find a tiny, frightened robot and his space capsule in their garden.

MR BERRY'S ICE CREAM PARLOUR

Jennifer Zabel

Carl is thrilled when Mr Berry, the new lodger comes to stay. But when Mr Berry announces his plan to open an ice-cream parlour, Carl can hardly believe it. And this is just the start of the excitements in store when Mr Berry walks through the door!

THE RAILWAY CAT AND DIGBY

Phyllis Arkle

Further adventures of Alfie, the Railway Cat, who always seems to be in Leading Railman Hack's bad books. Alfie is a smart cat, a lot smarter than many people think, and he would like to be friends with Hack. But when he tries to improve matters, by 'helping' Hack's dog, Digby, win a prize at the local show, the situation rapidly goes from bad to worse!

THE BUREAUCATS

Richard Adams

Imagine events in a large household, seen through the eyes of Richard and Thomas Kitten, who feel life should be organized entirely around them – it isn't of course, and the consequences are highly entertaining.

SEE YOU AT THE MATCH

Margaret Joy

Six delightful, stories about football. Whether they are spectators, players, winners or losers, these short, easy stories for young readers are a must for all football fans.

HAIRY AND SLUG

Margaret Joy

TV-mad Hairy, the Mablesdens' large, brown, shaggy dog, and Slug, the family's incredibly ramshackle little white car, have the most amazing adventures. The Mablesden family might think that they are going on a trip to the country, down to the shops or off to the carnival, but it's surprising how these ordinary every-day outings turn into something quite different when Hairy and Slug are around!

CHRIS AND THE DRAGON

Fay Simpson

Chris is always in trouble of one kind or another, but he does try extra hard to be good when he is chosen to play Joseph in the school nativity play. This hilarious story ends with a glorious celebration of the Chinese New Year.

HANK PRANK AND HOT HENRIETTA

Jules Older

Hank and his hot-tempered sister, Henrietta, are always getting themselves into trouble, but the doings of this terrible pair make for an entertaining series of adventures.

ON THE NIGHT WATCH

Hannah Cole

At the end of term the classrooms and playground will be locked and never used again. But no one is happy with the idea of sending all the children to different schools, and so teachers, parents and pupils get together to draw attention to their cause, all determined to keep their school open. Can the council be persuaded to change its mind? There is a very effective way of forcing them to listen . . .

ELOISE

Kay Thompson

At the Plaza Hotel, surrounded by her dog, her turtle, her nanny and a host of hotel guests, six-year-old Eloise is never bored . . .